*Children's Science Library*

# Scientists & Inventions

Author & Illustrator

**A.H. Hashmi**

Editor

**Rajiv Garg**

*Publishers*
**Pustak Mahal®**

*Published by:*

F-2/16, Ansari road, Daryaganj, New Delhi-110002
☎ 23240026, 23240027 • *Fax:* 011-23240028
*Email:* info@vspublishers.com • *Website:* www.vspublishers.com

**Branch : Hyderabad**
5-1-707/1, Brij Bhawan (Beside Central Bank of India Lane)
Bank Street, Koti Hyderabad - 500 095
☎ 040-24737290
*E-mail:* vspublishershyd@gmail.com

**Distributors :**

- **Pustak Mahal®**, Delhi
  J-3/16, Daryaganj, New Delhi-110002
  ☎ 23276539, 23272783, 23272784 • *Fax:* 011-23260518
  *E-mail:* sales@pustakmahal.com • *Website:* www.pustakmahal.com
  **Bengaluru:** ☎ 080-22234025 • *Telefax:* 080-22240209
  **Patna:** ☎ 0612-3294193 • *Telefax:* 0612-2302719
- **PM Publications**
  - 10-B, Netaji Subhash Marg, Daryaganj, New Delhi-110002
    ☎ 23268292, 23268293, 23279900 • *Fax:* 011-23280567
    *E-mail:* pmpublications@gmail.com
  - 6686, Khari Baoli, Delhi-110006
    ☎ 23944314, 23911979
- **Unicorn Books**
  **Mumbai :**
  23-25, Zaoba Wadi (Opp. VIP Showroom), Thakurdwar, Mumbai-400002
  ☎ 022-22010941 • *Telefax:* 022-22053387


ISBN 978-93-814483-9-7
**Edition 2011**



*Printed at :* Param Offset Okhla, Delhi

# CONTENTS

# SCIENTISTS AND INVENTIONS

It is very difficult to imagine the state of early man on earth when he had no home, no tools or weapons and no clothes. He had not learned to grow food for himself. He had to depend solely on his skill as a hunter and on gathering such edible fruits and nuts as he could find. In his search for food, man was different from the animals around him.

Man, as compared to other living beings, is endowed with a more active brain. It is only his brain which has made him far more progressive and advanced than any other animal.

Probably the first invention of man was a primitive tool made from a split stone which served a number of purposes. From this simple tool man developed the hand axe, knife and many other tools.

After this, roughly 500,000 years ago, man discovered how to make fire. Man had started using the skins of animals as clothes and living in caves. By 10,000 B.C. he had learned how to grow crops. Some 5,000 years ago, the wheel was invented which revolutionised transport. All these inventions resulted in the development of human civilisation.

The word 'Science' has been derived from Latin word 'Scientia' which means knowledge. Science is one of the most important subjects of study of our time.

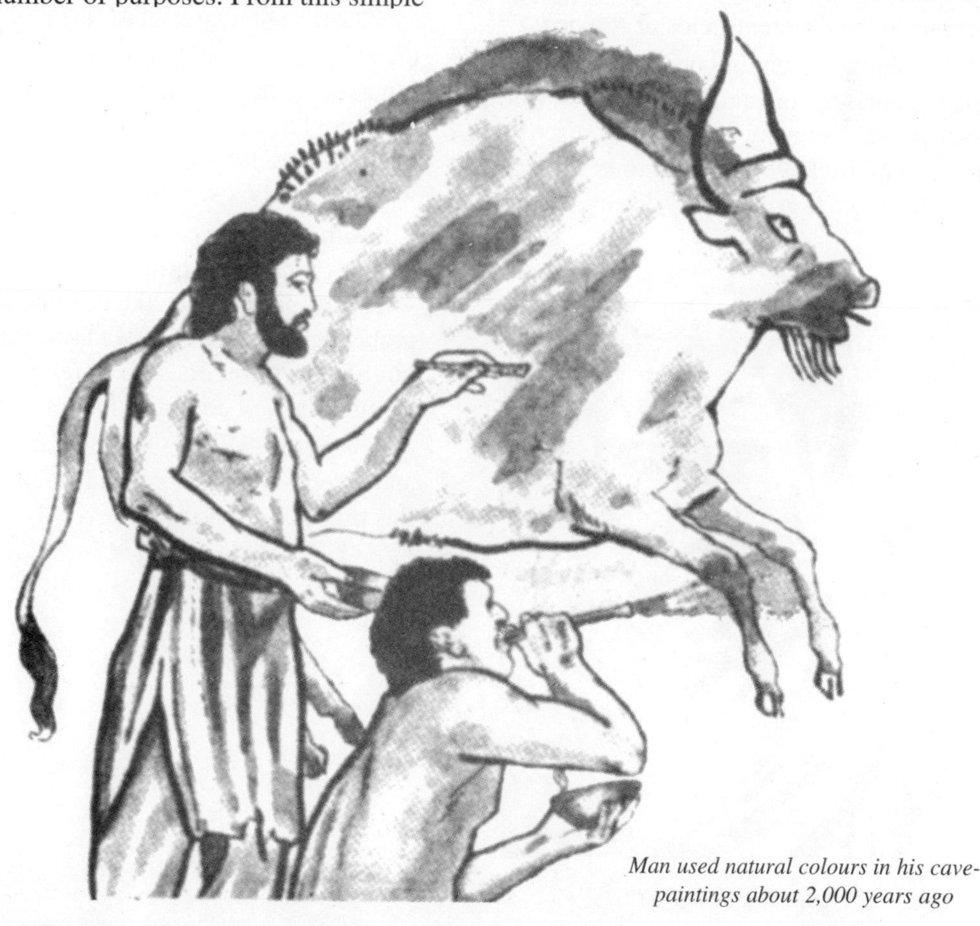

*Man used natural colours in his cave-paintings about 2,000 years ago*

The theories given by Pythagorus, Aristotle, Socrates, Plato, Archimedes, etc. have made significant contributions to scientific developments.

Alchemy – the ancient science of Egypt – contributed greatly in the development of Chemistry till the end of the 16th century. Charak, Sushrut of India, Jabir Ibn Hayyan, Al-Razi and Ibn-Sina, etc. were the famous scientists of that time.

The age of science really started in the middle of the 17th century. Robert Boyle for the first time introduced the method of experiments and its importance in the field of science. During this period, several institutions of science and scientists started working in England, France and Germany. Many scientists like Galileo, Newton, etc. brought forth many new principles and theories in the different fields of science.

Sometimes inventions occur all by accident. While working on cathode rays, Wilhelm Roentgen of Germany accidentally discovered the X-rays in 1895. The famous antibiotic Penicillin too was accidentally discovered by Sir Alexander Fleming. But such accidents occur rarely, and it needs an extra-intelligent brain to turn these chance-achievements into scientific discoveries. However, most of the inventions and discoveries are the result of a systematic research work. Sometimes necessity compels the scientists to discover new things. German scientists had to develop the technology of rockets and missiles to destroy England and the U.S. scientists invented atom bomb to defeat Japan in the Second World War. Radars and Sonars were necessary for self-defence. All these inventions are very useful for us in the present age.

*Aristotle (384-322 B.C.)*

Man succeeded in reaching the space only because of science. Computer has brought revolutionary changes in the field of information and telecommunication technology.

There is a great deal of difference between the words 'invention' and 'discovery'. A discovery is made of a thing that is already present in nature but is not known to man. An invention is a man-made device, for example, fire was discovered while match-box was invented.

Although science is beneficial for mankind, at times it can be harmful as well. Most of the scientists, prior to the 20th century, had not had proper schooling, but nowadays, we have well qualified engineers and scientists working together in well equipped laboratories.

From the very ancient times, the discoveries and inventions are being made in the world and this process will go on for ever.

# GALILEO GALILEI (1564–1642)

Galileo Galilei is one of the greatest scientists of the 16th and 17th centuries. He, with the help of his self-made telescope, noticed the mountains and valleys on the moon, the satellites of Jupiter and the rings of Saturn for the first time. He also proved that the milky way is a cluster of millions of stars.

Galileo was born on February 15, 1564 at Pisa (Italy). At the age of 23, he was appointed as a lecturer of mathematics. In 1581, he invented pendulum. Later his son developed wall clocks with the help of this pendulum. The development of barometer and steam engine was also an outcome of his inventions.

During this time the earth was considered to be the centre of the universe. Galileo, however, proved that earth is not the centre of the universe but like other planets it also revolves round the sun. Galileo was the first to prove Aristotle's statement wrong that if two balls of different masses are allowed to fall simultaneously from the same height, the heavy body will hit the ground first. For this he selected the leaning tower of Pisa and two metal balls weighing 100 pounds and 1 pound respectively. These balls were allowed to fall freely at the same time from the roof of the leaning tower of Pisa. Both the balls hit the ground at the same time, proving a myth wrong.

Old Galileo was imprisoned for blasphemy. After his release from the jail he became blind and died on January 8, 1642 in Italy. He collected his thoughts and theories in his famous book *Dialogues Concerning the Two Principal Systems of the World*. The great scientist Galileo also paved the way of space research for us.

*Galileo Galilei (1564-1642)*

# SIR ISAAC NEWTON (1642–1727)

Newton is famous for his law of Gravitation which he propounded after seeing an apple falling from the tree to the ground. Sir Isaac Newton was born on December 25, 1642 at Woolsthorpe, Lincolnshire. Even after 350 years, he is still considered as the Father of Physics.

As a child, Newton was not a very intelligent student. He did his graduation in 1665 from Cambridge University. In 1669, Newton was appointed as the professor of Mathematics in Trinity College. In 1672, Newton was elected the Fellow of the Royal Society of London. In 1689, Newton was elected the Member of Parliament and in 1703, he became the President of the Royal Society. Thereafter he was re-elected for the same post every year until his death. In 1705, he was knighted by Queen Anne in a special ceremony at Cambridge.

*Sir Isaac Newton (1642-1727)*

Newton propounded the Law of Gravitation and his three laws of motion which are still relevant and taught to the students of Physics. In 1687, he published a book entitled *The Mathematical Principles of Natural Philosophy*, in which the law of gravitation and laws of motion were explained. Newton showed that the white sunlight in fact is composed of seven colours namely, violet, indigo, blue, green, yellow, orange and red. His researches concerning light has been published in the book *Opticks*. He also invented the mathematical method known as Calculus.

Even in old age, Newton worked in the field of Astronomy. At the age of 85, he died on March 20, 1727. Even today Newton is considered as one of the greatest scientists of the world.

# SIR HUMPHRY DAVY (1778–1829)

Sir Humphry Davy is well known in the world for his invention of safety lamp for coal miners. He also discovered nitrous oxide, a well known anaesthetic gas known as laughing gas. He obtained this gas by heating ammonium nitrate. Apart from this he also developed the methods of analysing the compounds of nitrogen and oxygen and obtaining potassium nitroso sulphate. But his most important contribution was the use of electricity in the experiments of chemistry. He obtained several new elements by the method of electrolysis.

*Sir Humphry Davy (1778-1829)*

Sir Humphry Davy was born in Penzance (England) on December 17, 1778. He began his career with a medical practitioner but soon he acquired a good knowledge of natural sciences. On the basis of his knowledge, he was able to get a job in the Royal Institution. Here, he used to deliver popular lectures on the various topics of science. His fame spread far and wide and he was appointed a lecturer in the Royal Institution. He was made a knight in 1812 and later became Baronet. On May 29, 1829, he died at Geneva. He obtained sodium, potassium and some alkaline earth metals through electrolysis. He proved that chlorine is an element.

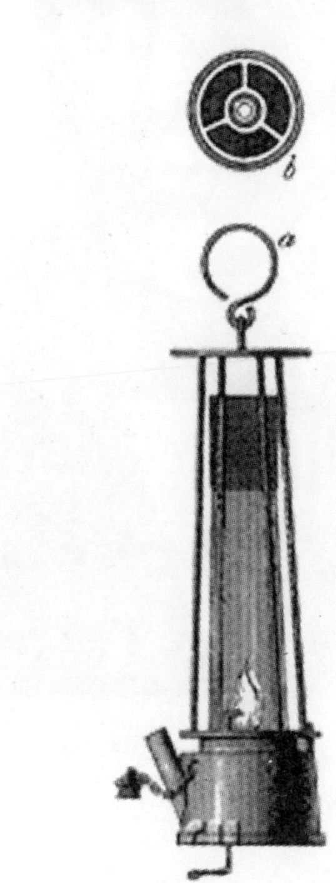


*The safety lamp invented by Davy*

In 1813, Faraday worked with him as his assistant and accompanied Davy on a tour of Europe during which they studied iodine and proved that diamond is an allotrope of carbon. Development of electric arc was also one of his inventions. He also established the catalytic action of red hot platinum on the mixture of inflammable gases. We cannot forget Davy's contributions in the field of science.

# MICHAEL FARADAY (1791–1867)

Michael Faraday is known as the father of Electromagnetic Induction. In 1831, he made the first dynamo to produce electricity. As a result of this invention, thousands of power stations are producing electricity all over the world. Without the invention of dynamo, production of electric power could not have been a reality. Faraday also made significant contributions in the field of electro-Chemistry.

*Michael Faraday (1791-1867)*

Michael Faraday, a British scientist, was born on September 22, 1791 at Newwington (England). His father was a blacksmith. At the age of 13, Faraday had to work as an apprentice in a book binding shop. He developed an interest in science after he attended some lectures of Sir Humphry Davy. He requested Davy to accept him as his assistant. In 1813, he began to work as Davy's assistant. In 1824, he became the member of the Royal Institution. Soon, he became the Director of the laboratory of the Royal Institution. In 1823, he was appointed as lecturer of Chemistry in the Royal Institution. In 1839, he fell seriously ill. After that his memory became quite weak. In 1861, after his retirement from the Royal Institution, he moved to a house in Hampton Court that was offered to him by Queen Victoria. On August 25, 1867, he breathed his last.

Most of the contributions made by Faraday are related with Physics. In Chemistry, his most important contribution is the 'laws of electrolysis'. He discovered paramagnetism and diamagnetism. He also proved that by high pressure, gases can be converted into liquid state. He also synthesized some chlorides of carbon. Two electrical units are named after him. One is Faraday unit that is used for measuring the quantity of electricity and the other is Farad unit, used to measure the capacity of a capacitor.

# CHARLES ROBERT DARWIN (1809–1882)

Charles Robert Darwin was the British Naturalist who is best known for his theory of evolution. He proved that all living things on earth have descended from the common ancestors who existed millions of years ago. All living beings – plants and animals – have evolved in an orderly way and continue to change even today.

Charles Darwin was born on February 12, 1809 at Shrewsbury. As a child, he was very much interested in studying and collecting insects and minerals. At the age of 16, he was sent to Edinburg University to study medicine but he had no interest in medicine. After that he was sent to Cambridge to study theology. At Cambridge, the professors of Botany and Geology became his good friends.

In 1831, a scientific expedition was being sent to sail around the world in HMBS Beagle under the command of Captain Fitz Roy. A naturalist was needed in this expedition for which Darwin's name was recommended by his two friends. He went on the voyage for five years and made collections of the bones of extinct animals and plants' remains. He surveyed the rain forests of Brazil. He found the fossil of an extinct giant animal Megatherium near Bahia Blanca. The strange marine iguanas, tortoises and the finches on the Galapagos Islands in the Pacific puzzled him at first because similar, yet quite different forms of the same animals appeared on separate islands.

*Darwin studied the iguanas and tortoises in the Galapagos islands*

*Charles Robert Darwin (1809-1882)*

During the same period, another naturalist, Alfred Russel Wallace also drew similar conclusions about evolution. In 1858, both Darwin and Wallace presented a joint paper. In 1859, the famous book of Darwin, *The Origin of Species* by Natural Selection was published. The first edition of the book was sold out on the day of its publication itself.

In 1868, his second book entitled *The Variation of Animals and Plants Under Domestication* was published. His other famous books are – *Insectivorous Plants, The Power of Movement in Plants* and *Descent of Man*.

Darwin died on April 19, 1882 at the age of 74 and was buried in Westminster Abbey near the tomb of Sir Isaac Newton.

# FRIEDRICH WOHLER (1800–1882)

The German scientist Friedrich Wohler discovered that it was possible to synthesize an organic compound urea from an inorganic chemical. It was one of his key contributions to revolutionise the concepts of organic and inorganic Chemistry.

Friedrich Wohler was born at Aschersheim near Frankfurt. He studied Chemistry under the guidance of Leopold Gmelin and Berzelius. In 1825, he became a lecturer at Berlin Technical school and then in 1831 in Casscel. In 1836, he became a lecturer of Chemistry at University of Gottingen. Here he assisted Liebig in many programmes. This outstanding scientist obtained cyanogen iodide in his student life itself. He demonstrated Pheroah's serpent action by burning mercuric thio-cyanide. He synthesized silver isocyanide and gave its chemical formula.

Most of the work of Wohler was carried out at Berlin. In 1827, he obtained aluminium metal by the reaction of potassium and aluminium chloride. This was a great achievement in the field of Chemistry. In 1828, he did some experiments with hydrocyanic acid and obtained the crystals of urea. Synthesis of urea in the laboratory was one of the greatest achievements because scientists like Berzelius had a concept that such compounds can only be obtained from nature and their synthesis in the laboratories is not possible. But Wohler, who was the student of Berzelius, could make it in the laboratory. In this way, he gave birth to Synthetic Organic Chemistry.

*Friedrich Wohler (1800-1882)*

In 1832, while he was working with Liebig, he extracted Bezoic acid from bitter almonds. He also carried out research work on quinone, hydroquinone and quinhydrone. He also worked on Alkaloids. He obtained phosphorus from bone ash and acetylene from calcium carbide. He also worked on hydrides of boron and silicon and on their rare compounds. This great chemist died in 1882.

# GREGOR JOHANN MENDEL (1822–1884)

Gregor Johann Mendel, the son of a farmer, was born on July 22, 1822 at Heinzendorf, Austria. He was a naturalist and studied science at the University of Vienna. After his education, he returned to his monastery at Bruno in 1847 and taught natural sciences in the school there.

For conducting experiments in genetics, Mendel grew pea plants in his monastery garden and did cross pollination between the different varieties. From 1856 to 1864 he conducted various experiments on pea plants on genetics and drew many conclusions regarding how one characteristic is related to the other. These conclusions were published in the magazine of Natural History Society. No scientist paid any attention to the research work done by Mendel for a period of 34 years. His work could be valued only around the end of 19th century when similar conclusions were drawn by the scientists of Holland, Germany and Austria.

After 16 years of his death on January 6, 1884, the laws of genetics propounded by Mendel could be understood by the scientists. According to him, some characteristics of the parents are inherited by the coming generations. These inherited characteristics are called hereditary characters. After this, Mendel became famous as father of genetics. He gave three laws – law of dominance, law of segregation and law of independent assortment, of genetics which are taught to the students even today.

*Gregor Johann Mendel (1822-1884)*

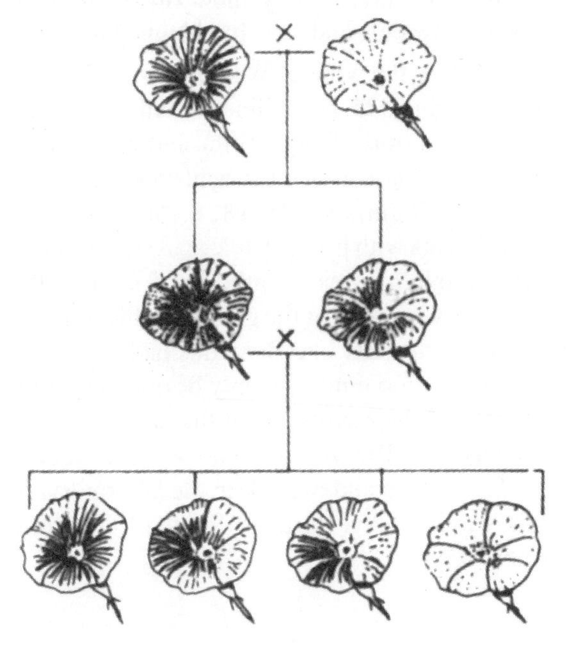

*He conducted various experiments on pea-plants*

# LOUIS PASTEUR (1822–1895)

Both science and mankind are indebted to the French chemist Louis Pasteur for his contribution in the development of science of Microbiology. He discovered the process of pasteurization to stop fermentation of milk and butter. He is well known for discovering the vaccine against the rabies.

*Louis Pasteur (1822-1895)*

Pasteur was born in Dole, France in 1822. He was the son of a poor tanner. After primary education, he went to Paris where he heard the lectures of Bellard and Duma. He became the assistant of Bellard and devoted himself to research work. His interest bent towards the study of crystals. In 1848, he was appointed as the lecturer of Physics at Lycee. In 1852, he became the lecturer of Chemistry at Strasberg. In 1857, he was appointed the Director of Ecole Normale, Paris. Here he worked on Fermentation by which originated a new branch of science called Microbiology. He invented rabies vaccination and experimented it on a nine-year-old boy whom a mad dog bit 14 times.

In 1888, the Pasteur institute was established and he remained its Head until his death.

# FRIEDRICH AUGUST KEKULE (1829–1896)

The German chemist, Friedrich August is well known all over the world for his famous discovery of the unique structure of the benzene molecule. Born in Darmstadt, Kekule was interested in the architecture and he planned a career as an architect. But his contact with Liebig changed his mind towards Chemistry.

After receiving his doctorate in 1852, he started his studies at Paris. In 1856, he was appointed as the lecturer of Chemistry at Heidelberg University. In 1858, he became the professor of Chemistry in Belgium at Ghent University. He moved to Bonn in 1865.

There is a famous story about the invention of the structure of benzene. It is said that on one night, in 1865, Kekule had a dream in which he saw the chains of the atoms of Carbon join together like snakes with each other's tails in their mouth. From this dream, he presented the ring structure of benzene.

On the basis of tetravalency of Carbon, he explained the Carbon bonds in molecules. He discovered that the atoms of Carbon can join onto each other in three different ways – in open chains, closed chains and ring chains. This discovery became so important that it has created more than 700,000 Carbon compounds to the present day.

Kekule in the 19th century brought revolutionary changes in writing the structural formulae of Carbon compounds. He called benzene and its compounds with a different name of aromatic compounds. This implies to all ring compounds. Hofman called other compounds as aliphatic compounds. The compounds of hydrogen and carbon are named as saturated and unsaturated hydrocarbons. Kekule, by adding ane, ene, and ine to the name of hydrocarbons classified them as alkanes, alkenes and alkines.

This revolutionary scientist of Chemistry died on July 13, 1896. His contributions in the field of science can never be forgotten.

*Friedrich August Kekule (1829-1896)*

H, C (benzene structural formula) ou

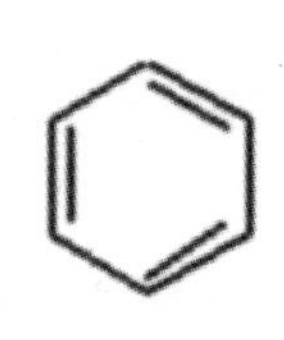

ou

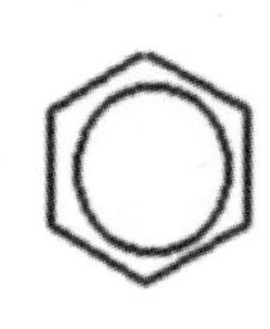

# THOMAS ALVA EDISON (1847-1931)

Thomas Alva Edison, during his life time took out more than 1,000 patents for his inventions. In the history of science perhaps there is no other scientist who has been credited with so many inventions and discoveries. Edison was born in Milan, USA on Feb. 11, 1847. When he was only 10 years old, he set up his first laboratory in the basement of his home. During 1860, he worked in the telegraph office of USA and Canada. In 1876, he sold telegraph printer he had invented. With this money, he could set up a good laboratory for his work at Menlo Park, New Jersey. In this laboratory he invented the Carbon-resistance transmitter and the phonograph. This was the first phonograph with which human voice could be recorded and reproduced.

In 1879, he invented the electric bulb. He also invented thermionic emission which is known as Edison effect today. In 1880, he lighted Menlo Park with 100 electric bulbs. This was a small demonstration of his invention. New York was the first city of the world where electric lights were used. Edison laid a cable of 75 km length and lighted more than one thousand houses on the night of September 4, 1882.

In 1887, he made a new laboratory in West Orange. Here in 1891, he invented Kinetograph. It was the first movie camera. In this camera, Eastman film was used. After developing the film, people used to see it with kinetoscope.

Edison was known as the wizard of technical age. He continued his scientific research until his last breath on October 18, 1931.

*Thomas Alva Edison (1847-1931) invented more than 1,000 things*

# ALBERT EINSTEIN (1879-1955)

Albert Einstein is known as the father of modern Physics. He was born on March 14, 1879 in Ulm, Germany. He became a Swiss citizen in 1901. In 1933, he moved to USA due to the fear of brutality of Hitler.

Right from his childhood, Einstein was interested in Science and Mathematics. His father was not very rich to bear the expenses of his education, so he had to earn along with his studies. When he was 15, his family moved to Italy. From Italy, he was sent to Switzerland for education. A rich relative helped him in getting higher education. Inspite of being highly intelligent, he could not get a job.

In Jurich University, his brilliance in Physics and Mathematics came to the surface. In 1900, he completed his education and became a citizen of Switzerland. After his education, he joined the Swiss patent office as a clerk. In his leisure, he used to solve the complex problems of Mathematics.

*Theory of Relativity is Einstein's major contribution to Science*

While working in the patent office, he published a paper which changed the concept of scientists about physical quantities. This research paper was on the Theory of Relativity. This paper brought worldwide fame for Einstein. In this paper he also gave the formula of $E = mc^2$ (E = energy, m = mass and c = velocity of light) regarding the conversion of mass into energy. This formula later became the basis of atom bomb which destroyed Hiroshima and Nagasaki in 1945. He felt deep sorrow at the destruction of these two cities by atomic explosions. After this, he spent his whole life on the peaceful uses of atomic energy.

Einstein gave the quantum theory of photoelectric effect for which he was awarded the Nobel Prize in 1921.

After retiring from Princeton University in 1845, Einstein continued his research on the unified field theory. He could not complete this work in his life time. On April 18, 1955, this great scientist died at Princeton Hospital. He died while sleeping. After his death, his brain was preserved in Princeton Hospital. To honour him, an element 'Einsteinium' has been named after him.

# ENRICO FERMI (1901-1954)

Born in Italy on September 29, 1901, Enrico Fermi was the first scientist to discover chain reactions. He made the first nuclear reactor of the world and marked the beginning of nuclear physics. In fact, he is known as the 'father of nuclear physics'.

*Enrico Fermi (1901-1954)*

In 1922, he obtained his Ph.D. degree in the field of X-rays from the University of Pisa. In 1927, he was appointed as a lecturer of Physics in the University of Rome. He was elected a member of Italian Academy in 1929. In 1933, he discovered a fundamental particle called neutrino for which he was awarded the Nobel Prize in Physics in 1938.

In those days, Mussolini's dictatorship caused a chaos in Italy. Fermi's wife was a Jew, so it directly affected him. Fortunately, Fermi was invited to give a lecture at the University of Columbia. Fermi, along with his family, went to USA and never came back to Italy. In 1939, he was appointed as professor of Physics in the University of Columbia and became a US citizen in 1944.

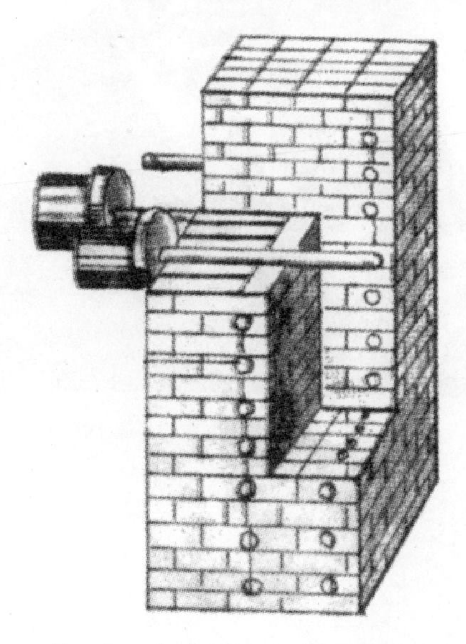

*Fermi made the first nuclear reactor*

At the University of Columbia, he started working on nuclear chain reactions. After a devoted hard work, he succeeded in making the first nuclear reactor in 1942. Fermi and his associates worked on the Manhattan project for making the atom bomb. In 1945, this team developed the atom bomb.

After World War II, Fermi joined the University of Chicago where the Institute for Nuclear Studies was named after him. He died on November 28, 1954 at the age of 53. In his honour, an element 'Fermium' has been named after him.

# J. ROBERT OPPENHEIMER (1904-1967)

J. Robert Oppenheimer was a successful administrator, a well known physicist, mathematician, researcher and an educationist. He is known to the whole world as the first director of the Manhattan project for developing the first atom bomb. A team of scientists started working together to develop the first atom bomb in 1943 at Alamos and successfully completed and tested it on July 16, 1945 at 5.30 a.m. On August 6, 1945, the first atom bomb was dropped at Hiroshima, Japan which killed 80,000 people and injured about 70,000. On August 9, 1945, the second bomb was dropped at Nagasaki, another city of Japan, which killed 40,000 people and injured 25,000. These inhuman acts of devastation gave a big jolt to Oppenheimer and he resigned from his post.

*J. Robert Oppenheimer (1904-1967)*

Oppenheimer was born in New York on April 22, 1904. His parents were rich Jews of Germany who had settled in the USA.

In 1925, he graduated in Physics from the Harvard University. After this, he joined Gavendish laboratory, England and worked with Ernest Rutherford.

In 1929, he came to the University of California and was appointed Associate Professor in 1931. Here he did many experiments in nuclear physics.

In 1941, President Roosevelt sanctioned the Manhattan project for making the atom bomb. A team of scientists was selected for this work and Oppenheimer was appointed the chief director of the project.

After World War II, Oppenheimer resigned from his post but again in 1947, he was made the chairman of the US Atomic Energy Commission. In 1963, he was honoured by the Fermi Award for the Atomic Energy Commission. Oppenheimer died on February 18, 1967 at Princeton.

# DR. HOMI JAHANGIR BHABHA (1909-1966)

The credit of starting the nuclear energy programme in India goes to Dr. Homi Jahangir Bhabha. After constituting the Atomic Energy Commission of India in 1948, Dr. Bhabha was appointed its first chairman. Indian scientists worked for the development of atomic energy under the able guidance of Bhabha. He was not only a well known scientist but also a successful administrator. He is known as Oppenheimer of India.

In 1956, the first atomic reactor, Apsara was installed under his supervision. Two other nuclear reactors, Cirus and Zerlina were also installed under his guidance. It was the result of Bhabha's efforts that the first nuclear power plant for the production of electricity came into existence at Tarapur. Two years later, a plutonium plant was made and on May 18, 1974, an underground atomic explosion test at Pokharan was conducted. This was the result of the efforts made by Dr. Bhabha.

Dr. Homi Bhabha was born on October 30, 1909 in Bombay (now Mumbai), in a wealthy Parsi family. His primary education was completed in Bombay (now Mumbai). After graduation, he was sent to the Cambridge University for higher education. From Cambridge, he obtained an engineering degree in 1930 and a Ph.D. in 1934. Here he worked with a world renowned scientist, Niels Bohr. He also worked with Enrico Fermi and Pauli.

*Dr Homi Jahangir Bhabha (1909-1966)*

Dr. Bhabha worked a lot on cosmic rays. In 1945, he established the Tata Institute of Fundamental Research and became its director. He was the chairman of the first United Nations' Conference on Peaceful Uses of Atomic Energy held at Geneva in 1955. Dr. Bhabha died in an air crash on January 24, 1966 when he was on his way to attend an International conference.

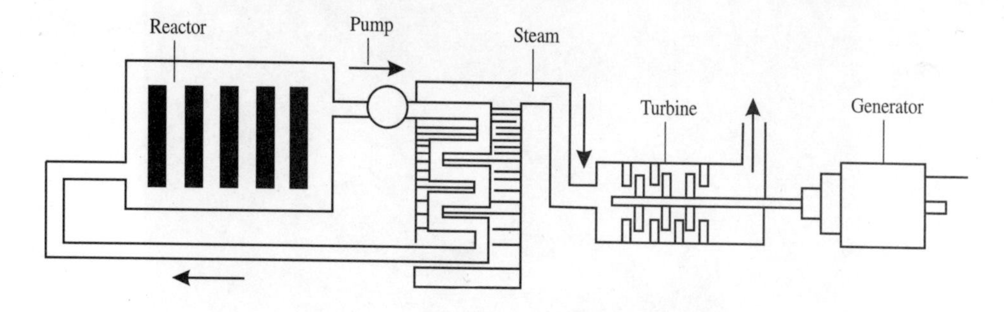

# GREAT INVENTIONS

The development of human civilisation is a result of inventions made by men. Metal smelting started about 4000 B.C. which was one of the greatest ancient inventions. The wheel and plough came to be known around 3000 B.C. The speed of inventions was slow upto 14th century but from the 17th century, science developed rapidly. Some great inventions are listed below:

| | |
|---|---|
| 1450 | Printing Press, Johannes Gutenberg (Germany) |
| 1590 | Compound Microscope, Zacharias Janssen (Netherlands) |
| 1593 | Thermometer, Galileo Galilei, (Italy) |
| 1608/09 | Refracting T,elescope, Hans Lippershey (Netherlands); Galileo Galilei (Italy) |
| 1668 | Reflecting Telescope, Isaac Newton (Britain) |
| 1698 | Steam pump, Thomas Savery (Britain) |

*Hot air balloon made by Montgolfier Brothers*

| | |
|---|---|
| 1712 | Beam Engine, Thomas Newcomen (Britain) |
| 1733 | Flying Shuttle, John Kay (Britain) |
| 1767 | Spinning Jenny, James Hargreaves (Britain) |
| 1783 | Hot Air Balloon, Montgolfier Brothers (France) |
| 1784 | Improved Steam Engine, James Watt (Britain) |
| 1785 | Power Loom, Edmund Cartwright (Britain) |
| 1792 | Cotton Gin, Eli Whitney (USA) |
| 1800 | Electric Battery, Allessandro Volta (Italy) |
| 1800 | Lathe, Henry Maudslay (Britain) |
| 1804 | Steam Locomotive, Richard Trevithick (Britain) |
| 1815 | Saftey lamp, Humphry Davy (Britain) |
| 1815 | Stethoscope, Rene T.H. Laenec (France) |
| 1836 | Revolver, Samuel Colt (USA) |
| 1837 | Telegraph, William Cooke and Charles Wheatstone (Britain); Samuel Morse (USA) |
| 1839 | Steam Hammer, James Nasmyth (Britain) |
| 1845 | Sewing Machine, Elias Howe (USA) |

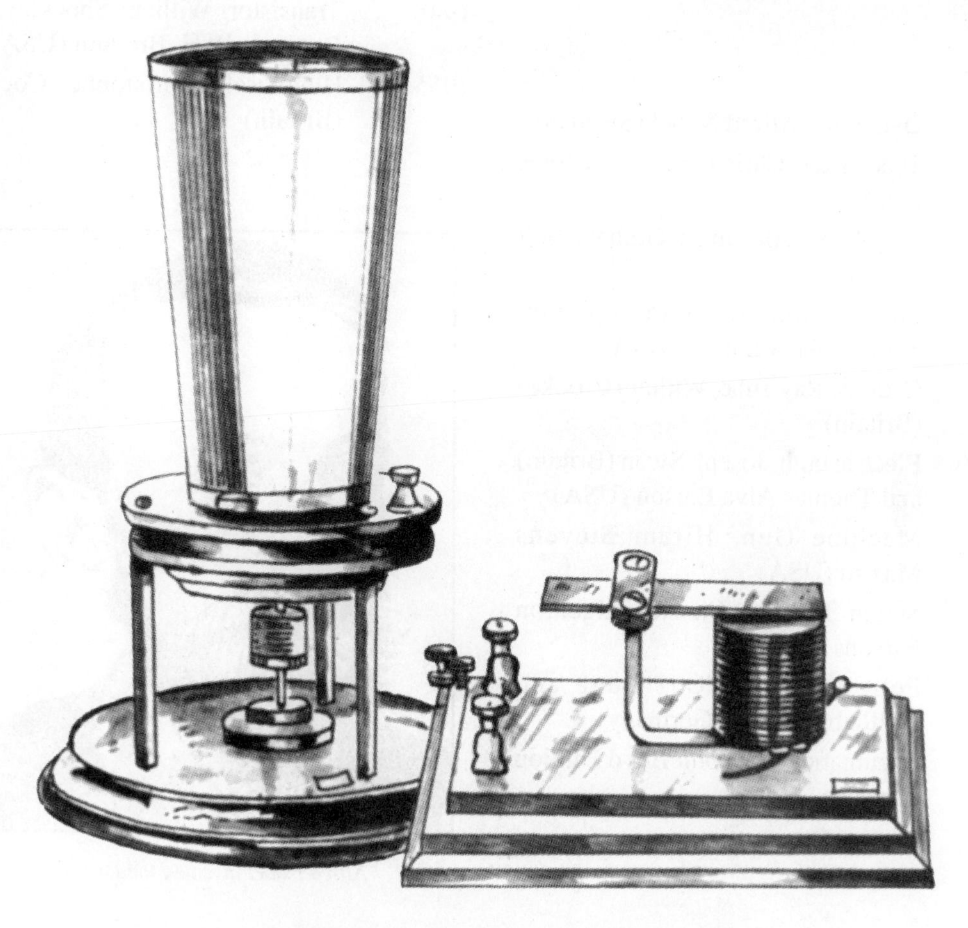

*Alexander Graham Bell invented Telephone in 1876*

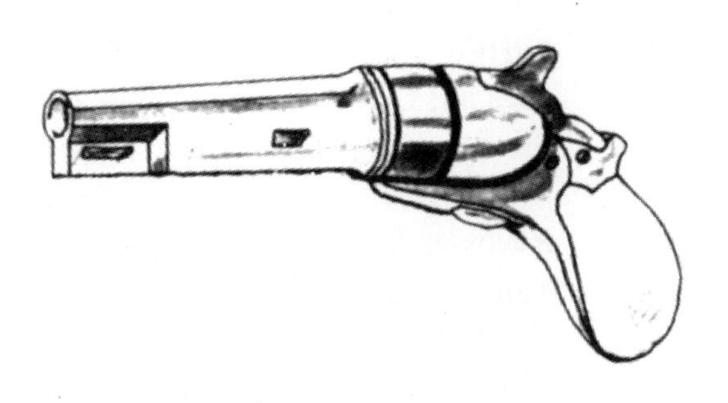

*Samuel Colt made the first revolver*

| | |
|---|---|
| 1867 | Dynamite, Alfred Nobel (Sweden) |
| 1872 | Typewriter, Christopher L. Scholes (USA) |
| 1876 | Telephone, Alexander Graham Bell (USA) |
| 1877 | Phonograph or Gramophone, Thomas Alva Edison (USA) |
| 1878 | Cathode Ray Tube, William Crookes (Britain) |
| 1878/79 | Electric bulb, Joseph Swan (Britain) and Thomas Alva Edison (USA) |
| 1880 | Machine Gun, Hiram Stevens Maxim (USA) |
| 1884 | Steam Turbine, Charles Algemon Parsons (Britain) |
| 1885 | Petrol Engine, Karl Benz and Gottlieb Daimler (Germany) |
| 1888 | Pneumatic Tyre, John Boyd Dunlop (Britain) |
| 1893 | Diesel Engine, Rudolf Diesel (Germany) |
| 1895 | Radio, Guglielmo Marconi (Italy) |
| 1903 | Aircraft, Wilbur and Orvile Wright (USA) |
| 1926 | Television, John Logie Baird (Britain) and Vladimir Zworykin (USA) |
| 1930 | Cyclotron, Ernest Lawrence (USA) |
| 1937 | Jet Engine, Frank Whittle (Britain) |
| 1944 | Digital Computer, Howard Aiken (USA) |
| 1947 | Polaroid Camera, Edwin H. Land (USA) |
| 1948 | Transistor, William Shockley, John Bardeen, W.H. Brattain (USA) |
| 1955 | Hovercraft, Christopher Cockerell (Britain) |

*Alfred Nobel invented the Dynamite*

| | |
|---|---|
| 1971 | Micro Processor, Intel Corporation (USA) |
| 1972 | Commercial Video Games, Pocket Calculator and Home Video System. |
| 1975 | Body Scanners. |
| 1977 | Apple II, Personal Computer. |
| 1978 | First Test Tube Baby. |
| 1980 | Compact Disk Audio System. |
| 1985 | Compact Disk Memory System. |
| 1986 | Challenger Space Shuttle. |
| 1987 | Docklands Light Railway (Britain) |
| 1990 | High-Definition TV System. |

■■

*Edison invented Gramophone in 1877*

# SCIENCE AND TECHNOLOGY

**Measurement of Time:** Sundial and water clocks were the first instruments discovered to measure time in about 1500 B.C. In the sundial, the sun casts the shadow on the digits and gives the indication of time. In the water clock, water from one container falls drop by drop into another container. The container has an hour scale by which time is measured.

The first mechanical wall clock was developed in 1088 in China. It was 10 meters high and was run by water power. The first mechanical clock in Europe was made in 1200. The first clock was made in Spain in 1276. The oldest mechanical clock which is still working is in the Salisbury Cathedral. It was made in 1368.

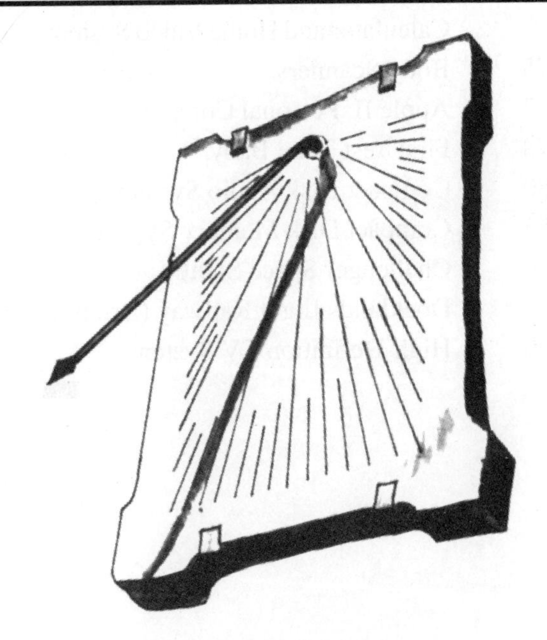

*The Sundial*

**Glass Manufacturing:** Glass was first made by melting soda and sand in about 3000 B.C. in Syria and its neighbouring countries. Glass blowing was also invented in Syria in about 100 B.C.

**First Balance:** The first weighing balance was developed in Syria and its neighbouring countries between 4000 and 5000 B.C. This balance was used to weigh gold. Stone weights were used to measure weight. These were cut in the shape of animals.

**Gunpowder:** Gunpowder was probably first made in China or India by mixing Sulphur, Charcoal and Saltpetre. Around 850 A.D., gun-powder was used by the Chinese for making fireworks and explosives. Gunpowder came in use during the 13th century in Europe. The credit of inventing gunpowder in Europe goes to an English monk named Roger Bacon.

*Glass was first made in Syria by melting soda and sand in about 3000 B.C.*

**Invention of Spectacles:** Impaired sight has always been a challenge to the scientists. As long ago as the year 1000, an Arab scientist Alhazen demonstrated the action of image formation by lenses. He also told that people who had weak eyesight could see properly with the help of lenses. During the year 1200, Roger Bacon made a pair of rudimentary spectacles. By 1430, Italians developed spectacles to view the distant objects clearly. Bifocal lenses were invented in the 18th century by American statesman, Benjamin Franklin.

**Production of Petrol:** Oil was extracted by drilling in the USA for the first time in 1841 and the oil well was made in 1859. Petrol was obtained from the crude oil in 1864. Petrol had not much utility before the invention of motor-car. After the invention of the motor-car in 1883, the first petrol station was opened in France in 1895. The first petrol refinery was developed in 1860 in the USA. In 1870, Standard Oil Company, the biggest in the world of that time, was established. In 1890, the high quality petrol production started.

**Steam Engine:** The first successful steam engine was made in 1712 by the British engineer, Thomas Newcomen. It was used to draw water out of the mines. James Watt modified the Newcomen engine. In 1765, he made a new type of steam engine which was more powerful and fast. Steam engines of Watt were used for the first time in 1785 to run cotton mills. These steam engines proved very useful in the industrial development. In the 19th century, the steam engines came to be widely used in road and water vehicles. In 1803, the locomotive was invented in Britain. The first successful rail engine was made by George Stephenson in 1814.

**Spinning Machine:** Spinning machine was invented in 1700 in Britain. Before this, yarn was made either by hand or by Charkha. The first spinning machine was Spinning Jenny invented by James Hargreaves in 1764. It was a hand-operated machine. This could spin very thin yarn. Another spinning machine was Arkwright's Water Frame which was made in 1769. Samuel Crompton combined both these machines and made a new machine named spinning mule which marked the beginning of the textile industry.

The first automatic spinning machine was made in 1801 in France. This was known as

*The first oil well was constructed in 1839*

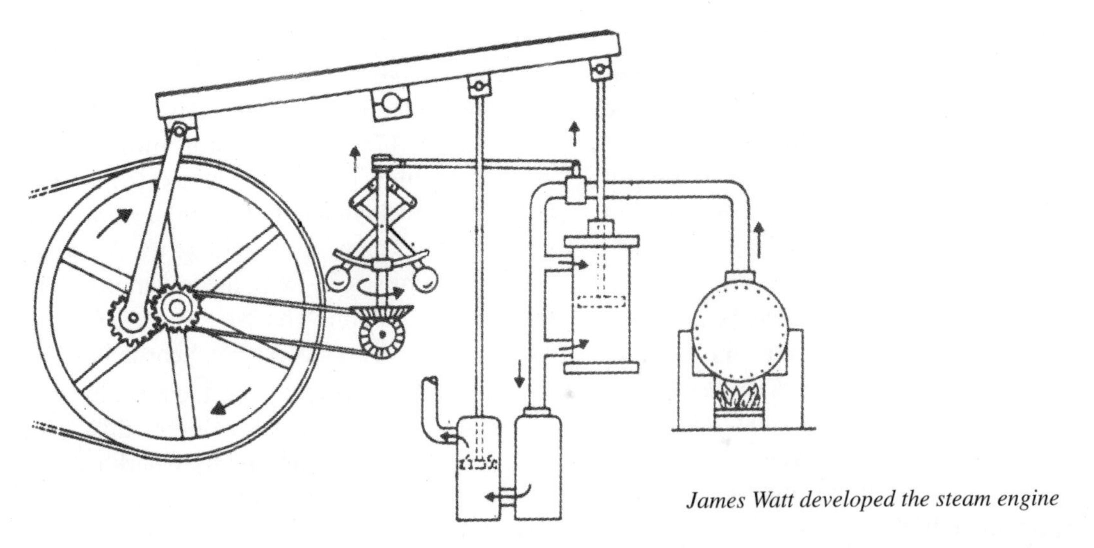

*James Watt developed the steam engine*

Loom and was capable of creating designs of silk clothes. It was invented by Joseph Marie Jacquard. It used a set of punched cards for creating new patterns. Nowadays computers have replaced punched cards.

**Sewing Machine:** The first sewing machine was made in 1830 in France by Barthelemy Thimmonier. This machine was able to put 200 stitches in one minute. The first successful

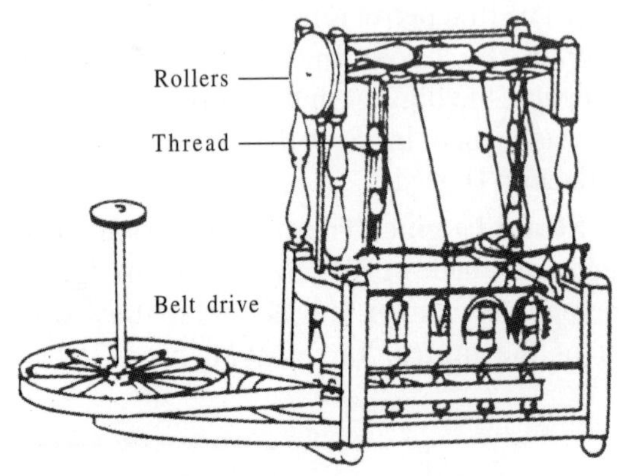


*Arkwright's water-frame machine*

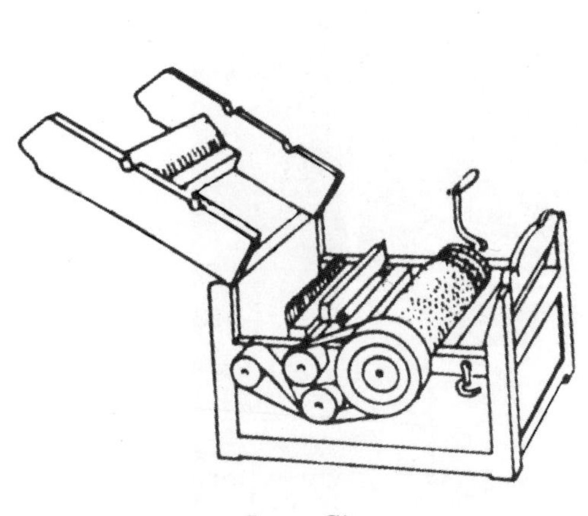

*Cotton Gin*

*Spinning mule*

sewing machine was made by Elias Howe of USA in 1845. Isaac Singer of USA gave the modern shape to the sewing machines in 1851.

**Harvesting Machines:** Harvesting machines are of two types: one is used for cutting the crops and the other for separating grains and fodder. The first threshing machine was developed in Britain in 1786 by Andrew Meikle. The first reaping machine was also invented in Britain in 1826 by Patrick Bell. The first harvesting machine was made in USA in 1831 by Cyrus McCormick. The maximum development work on harvesting machines has been done in the USA. Today these machines are being widely used all over the world.

**Invention of Electric Motor:** The first electric motor was invented by Michael Faraday in 1821. This was only an experimental motor. The first successful dynamo was made in Belgium in 1870 by Zenobe Theophile Gramme. After this, the first practical electric motor was made by him in 1873. The AC motor was invented by Nikola Tesla in 1888 in the United States of America.

**Electric Light:** Electric light for the first time was produced by an electric arc by Humphry Davy in Britain in 1802. Its light was very intense, so it was not good for domestic purposes. J.W. Starr and Joseph Swan tried to make an electric bulb but they did not succeed. The first successful electric bulb was made by the famous inventor, Thomas Alva Edison. Wire filament bulbs came into existence only in 1898.

**Artificial Dye:** The first artificial dye was made in 1856 by the British scientist, William Perkin. Prior to this, all dyes were made from insects and plants. Different artificial dyes have been made only after the invention of Perkin.

**Invention of X-rays:** X-rays were invented by the German physicist, Wilhelm Roentgen in 1895. These rays were invented accidentally while he was doing some experiments on Cathode rays. Today, X-rays are not only used to locate dislocations and fractures of bones but also in

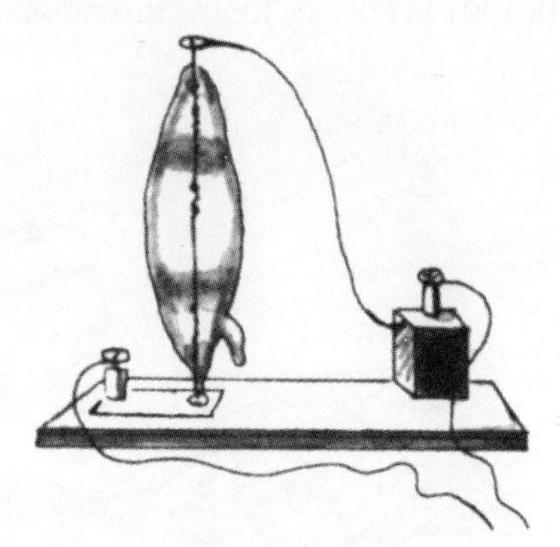

*Swan's electric bulb*

*Electric bulb made by Edison*

industries. Roentgen named them X-rays because these were not known at that time ('X' means unknown). Roentgen was given the first Nobel Prize in 1901 in Physics for the invention of X-rays.

**Invention of Plastic:** Plastic was invented for the first time by the British scientist, Alexander Parkes. This was called Parkesine and was made with cellulose and camphor. An

*Harvesting machine*

*William Perkin invented the artificial dyes*

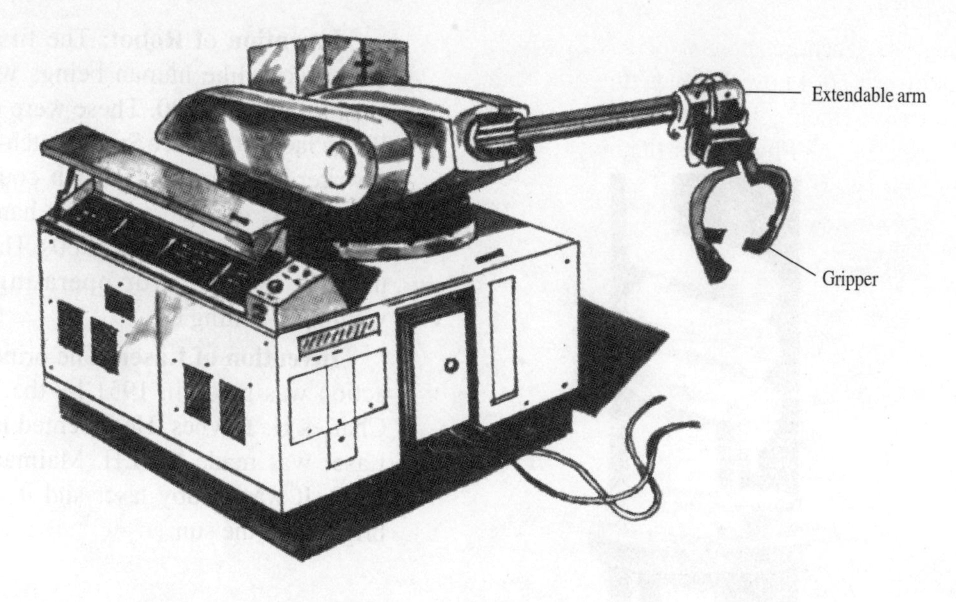


*Industrial robots*

American inventor, John Hyatt made similar plastic in 1868 which was named as celluloid. The first plastic made of chemicals was Bakelite which was invented by Leo Backeland of Belgium in 1907.

**First Telephone:** The telephone was invented by Alexander Graham Bell of USA in 1876. The telephone was first used in 1877 in Boston and first public call box was made in Connecticut in 1880. The first automatic telephone exchange was opened in 1892 at La Porte of Indiana. The automatic telephone exchanges came to be used in Europe in 1909.

**Beginning of Radio Broadcasting:** Radio waves were discovered in 1887 by Heinrich Hertz of Germany. The first signal in Morse Code was transmitted by Guglielmo Marconi of Italy in 1895. The first radio broadcast with music and talks was done by Canadian Reginald Fessenden on December 24, 1906 in USA. The first radio station was established in New York in 1907.

**Invention of Television:** The first television signal was transmitted by British inventor, John Logie Baird in 1924. This electronic system was different from the one being used today. The electronic television was developed in USA by Philo Farnsworth. Zworykin achieved a great success in 1930 in developing the electronic television.

**Nuclear Power:** Nuclear energy was first produced in 1942 by the famous Italian scientist, Enrico Fermi in USA. He made the first successful nuclear reactor in Chicago. In this reactor, Uranium was used as fuel. These days such reactors are used for electric power generation.

**First Computer:** The first computer called Colossus was developed in Britain in 1943. This could very quickly decode the codes of war. The first successful computer was made in USA in 1946. It was called ENIAC. About 19,000 valves and several thousand other electronic components were used in it. It was of the size of a big room.

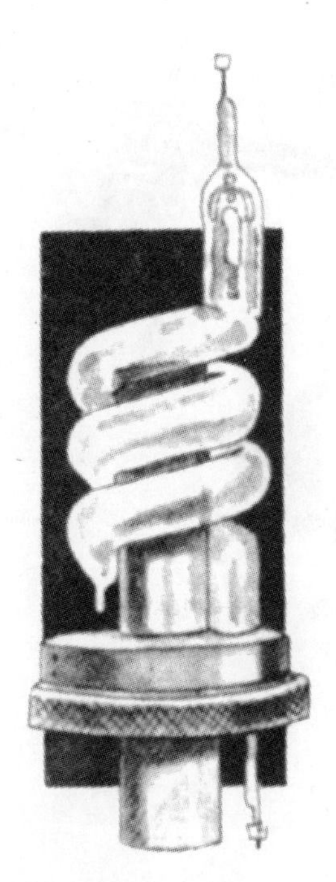

*Maiman invented laser-rays*

**Invention of Robot:** The first robots that could work like human beings were made in Europe around 1700. These were used as toys. Pierre Jacquet-Droz, a Swiss watch-maker, made a writer robot in 1770 which could write any message of 40 letters with its hand. Industrial robots were developed in 1960s. These are being used in factories for operating machines, welding, painting etc.

**Invention of Laser:** The principle of laser action was given in 1951 by the US scientist, Charles H. Townes. He invented laser in 1953. Laser was made by T.H. Maiman of USA in 1960. It was a ruby laser and it was far more bright than the sun.